THE GALAPAGOS ISLANDS

ISLANDS IN THE SEA

William Russell

The Rourke Book Co., Inc.
Vero Beach, Florida 32964

© 1994 The Rourke Book Co., Inc.

All rights reserved. No part of this book may be reproduced or utilized in any form or by any means, electronic or mechanical including photocopying, recording or by any information storage and retrieval system without permission in writing from the publisher.

Edited by Sandra A. Robinson

PHOTO CREDITS
All photos © James P. Rowan

Library of Congress Cataloging-in-Publication Data

Russell, William, 1942-
   The Galapagos Islands / by William Russell.
     p. cm. — (Islands in the sea)
   Includes index.
   ISBN 1-55916-031-4
   1. Zoology—Galapagos Islands—Juvenile literature.
2. Galapagos Islands—Juvenile literature. [1. Zoology—
Galapagos Islands. 2. Galapagos Islands.] I. Title. II. Series.
QL345.G2R87 1994
591.91866'5—dc20   93-48335
                                                     CIP
                                                     AC

TABLE OF CONTENTS

The Galapagos Islands	5
The Volcanic Galapagos	6
The Galapagos Long Ago	9
Strange Creatures of the Galapagos	11
Giant Tortoises	14
Lizards and Crabs	16
Seals and Sea Lions	19
Boobies and Other Birds	20
Map of the Galapagos Islands	21
Visiting the Galapagos	22
Glossary	23
Index	24

THE GALAPAGOS ISLANDS

The Galapagos Islands are one of nature's most remarkable creations. Old-time sailors called them the "Enchanted Isles."

Of course, the sailors did not find wizards there, 600 miles off the coast of Ecuador. They did, however, find big *lizards*. To the sailors' surprise, some of those lizards swam in the sea!

The men also found giant **tortoises** and dozens of other unusual animals.

A marine iguana, the lizard of the sea, naps on the volcanic rock of Fernandina Island in the Galapagos

THE VOLCANIC GALAPAGOS

The five large Galapagos Islands and the more than 60 smaller islands are the upper parts of **ancient,** or very old, **volcanoes.** Some of the volcano peaks rise more than 4,000 feet above the Pacific Ocean's surface.

Plants can grow on only part of the islands' volcanic ash and rock. Many are desert plants, like cactus.

An opuntia cactus towers above a marine iguana on South Plaza Island

THE GALAPAGOS LONG AGO

Explorers who sailed from Spain found the Galapagos Islands in the early 1500s. No one paid much attention to the islands for the next 300 years. Then whaling ships began to stop — and "shop"— at the islands. Their crews took water to drink and tortoises to eat.

Pirates and scientists came to the Galapagos, too. No one found the islands more interesting than Charles Darwin, a famous English scientist. In 1835, Darwin studied animals on 15 of the Galapagos Islands, and later he wrote about them.

Charles Darwin came to the Galapagos Islands in 1835 and studied the weird, wonderful animals — like this frigate bird

STRANGE CREATURES OF THE GALAPAGOS

For Darwin, the Galapagos Islands were like a zoo without fences. The animals had no fear of people. Most Galapagos animals are still not afraid.

One strange and unusual Galapagos creature is a bird — the flightless cormorant. It has wings, but they are too weak for flight.

The flightless cormorant, lava gull, Galapagos penguin and several other kinds of animals live there, but nowhere else in the world.

Flamingoes, penguins, albatrosses, iguanas, sea lions and many other animals also live on the islands.

The flightless cormorant is one of several kinds of Galapagos animals that live nowhere else in the world

On Hood Island, blue-footed boobies strike a courtship pose to attract mates

A Galapagos penguin waddles over the volcanic beach of Isabela Island

GIANT TORTOISES

Galapagos tortoises are plant-eating giants! A big one can weigh over 500 pounds.

In the early 1800s, sailors carried live tortoises onto their ships. Tortoises could live for months on the ships. When they wanted fresh meat, the sailors butchered a tortoise.

Today, Galapagos tortoises are protected from hunters, but they are rare on most of the islands.

A vermilion flycatcher hitches a ride on a giant Galapagos tortoise

LIZARDS AND CRABS

The Galapagos Islands have many crusty creatures. None is more unusual than the five-foot-long **marine** iguana. Unlike its land-loving cousins, this one-of-a-kind lizard swims in the ocean and feasts on seaweed.

Marine iguanas **bask,** or sun themselves, on seaside rocks. They share the rocks with armies of tomato-red Sally Lightfoot crabs and lava lizards.

Away from the sea, land iguanas munch on fruits and leaves.

Sally Lightfoot crabs add a dash of color to the dull lava rocks

SEALS AND SEA LIONS

The ocean waters around the Galapagos are rich with fish. The fish attract fur seals and sea lions. Seals and sea lions hunt in the ocean. They come ashore to bask, rest and have their young.

Galapagos fur seals are not common. In the 1800s, fur hunters killed thousands of them.

Sea lions are plentiful. Their fur had little value to fur hunters. Today visitors to the Galapagos **snorkel** with curious sea lions swimming around them like playful pups.

A Galapagos fur seal basks on the rocky shore of Santiago Island

BOOBIES AND OTHER BIRDS

The largest birds of the Galapagos Islands eat fish and other sea creatures. Boobies, flightless cormorants and Galapagos penguins dive for fish. Long-legged herons spear fish.

The albatross hunts while gliding on long, narrow wings over the sea. The frigate bird swoops to grab flying fish.

Mockingbirds and several kinds of finches are common land birds on the Galapagos.

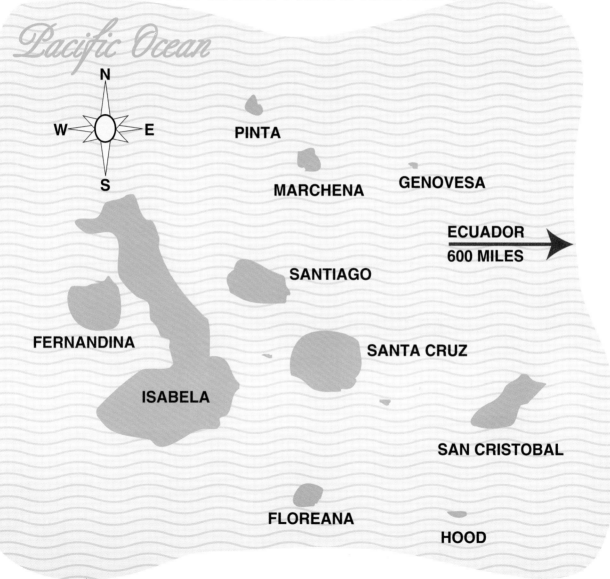

VISITING THE GALAPAGOS

The Galapagos Islands are a national park of Ecuador. Ecuador controls the number of visitors and where they can walk. Too many visitors hiking off paths could harm the islands' wildlife.

The islands are still recovering from old problems. Settlers and sailors brought goats, rats, mice, cats, dogs and farm animals to the islands. These animal pests destroyed many wild animals — such as tortoises — by either eating them or their food. Today, Ecuador is getting rid of as many animal pests as it can.

Glossary

ancient (AIN chent) — very old

bask (BASK) — to warm oneself by lying in sunshine

marine (muh REEN) — of or relating to the sea, salt water

snorkel (SNOR kul) — to swim with the help of a special breathing tube; the special breathing tube used by swimmers

tortoise (TOR tihs) — a turtle at home on land

volcano (vahl KAY no) — an opening in the Earth caused by underground forces, and the mountain of rock that forms around it

INDEX

albatross 11, 20
animals 5, 9, 11, 22
boobies 20
cormorant, flightless 11, 20
crabs, Sally Lightfoot 16
Darwin, Charles 9, 11
Ecuador 5, 22
explorers 9
finches 20
flamingoes 11
frigate bird 20
gull, lava 11
herons 20
iguana
   land 16
   marine 16
lizards 5
   lava 16

map of the
   Galapagos Islands 21
mockingbirds 20
Pacific Ocean 6
penguins 11, 20
pirates 9
plants 6
sailors 5, 14, 22
scientists 9
seal, fur 19
sea lions 11, 19
settlers 22
tortoises 5, 9, 14, 22
visitors 19, 22
whaling ships 9
wildlife 22
volcanoes 6